ANIMAL GENIUSES

ELEPHANTS

THERESA EMMINIZER

PowerKiDS press

Published in 2026 by The Rosen Publishing Group, Inc.
2544 Clinton Street, Buffalo, NY 14224

First Edition

Editor: Theresa Emminizer
Book Design: Michael Flynn

Photo Credits: Cover, pp. 1, 5, 9 Henk Bogaard/Shutterstock.com; (series background) Viktoriia LiSa/Shutterstock.com; p. 7 (African Savanna) Piotr Poznan/Shutterstock.com; p. 7 (African Forest) Roger de la Harpe/Shutterstock.com; p. 7 (Asian) Sourabh Bharti/Shutterstock.com; p. 11 John Michael Vosloo/Shutterstock.com; p. 13 Nora Yusuf/Shutterstock.com; p. 15 Michael K. McDermott/Shutterstock.com; p. 17 Nellariel/Shutterstock.com; p. 19 Vladimir Wrangel/Shutterstock.com; p. 21 givaga/Shutterstock.com.

Cataloging-in-Publication Data

Names: Emminizer, Theresa.
Title: Elephants / Theresa Emminizer.
Description: Buffalo, NY : PowerKids Press, 2026. | Series: Animal geniuses | Includes glossary and index.
Identifiers: ISBN 9781499451887 (pbk.) | ISBN 9781499451894 (library bound) | ISBN 9781499451900 (ebook)
Subjects: LCSH: Elephants–Juvenile literature. | Elephants–Behavior–Juvenile literature.| Animal intelligence–Juvenile literature. | Elephants–Psychology–Juvenile literature.
Classification: LCC QL737.P98 E45 2026 | DDC 599.67–dc23

Manufactured in China

CONTENTS

Excellent Elephants

Elephants are big, beautiful animals. Most live for a long time and grow throughout their life. They measure 10 feet (3 m) tall and weigh thousands of pounds. Elephants are well known for having great **memories**. They're also super smart! In this book, you'll find out what makes elephants so special.

An Elephant Parade

There are three species, or kinds, of elephants: African savanna (bush), African forest, and Asian. You can tell the species apart by the shapes of their ears and trunks. African savanna elephants are the largest land animals on Earth! Elephants live in a group called a herd or a parade.

African savanna

African forest

Asian

Herd Life

The oldest female elephant leads the herd. She is called the matriarch. The matriarch knows where to find food, water, and places to live. She also knows how to keep safe from predators. She helps guide the other females and young. Adult male elephants usually live on their own.

Working Together

Baby elephants are called calves. Calves drink milk from their mother and stay close to her until age 3. The herd works together to raise calves and keep them safe from harm. If a predator is near, the herd forms a circle around the calves, hiding them in the middle.

An Elephant Never Forgets

Elephants have very strong memories. That's because their temporal lobe, the part of the brain that holds onto memories, is bigger than most other animals'. An elephant's brain weighs 10.5 pounds (4.8 kg)! Elephants use their big brains to guide them to food and water, sometimes across great **distances**.

Elephant Talk

Elephants have lots of ways to **communicate** with one another. They may touch each other with their trunks, make loud calls, or other sounds. Some of these sounds are too low for people to hear, but elephants can hear them, even from very far away!

Elephant Empathy

Empathy is being **aware** of and sharing someone else's feelings. Elephants are very empathic animals. They comfort, or care for, one another when they're sad and help each other when hurt. Elephants **grieve** when a member of the herd dies and cover their body with leaves.

Crafty and Clever

Elephants find clever, or smart, ways to figure out problems. They use rocks and branches as tools. They may pile up rocks to stand on to reach food that's too high. They also play games and use toys! Elephants understand pointing and can **recognize** themselves in **mirrors**.

Help Elephants

Scientists are still learning about elephants' minds. Sadly, elephants have been put in danger by poaching (unlawful hunting) and **habitat** loss. These smart, special animals are in danger of going extinct, or dying out. Keeping elephants' wild homes safe will be a key part of making sure they live on.

GLOSSARY

aware: Knowing about something.

communicate: To share ideas and feelings through sounds and motions.

distance: The space between two points.

grieve: To express sadness for the loss of something or someone.

habitat: The natural home for plants, animals, and other living things.

memory: Thoughts of past people, places, or events.

mirror: A shiny surface that reflects an image.

recognize: To see and know something or someone.

FOR MORE INFORMATION

BOOKS

Markovics, Joyce L. *Elephants*. Ann Arbor, MI: Cherry Lake Publishing, 2023.

Parkin, Michelle. *A Memory of Elephants*. Parker, CO: A Child's World, 2024.

WEBSITES

National Geographic Kids: 10 Unforgettable Elephant Facts!
www.natgeokids.com/uk/discover/animals/general-animals/elephant-facts/
Find out more awesome facts about elephants!

San Diego Zoo: Elephant
animals.sandiegozoo.org/animals/elephant
Learn more about different elephant species.

INDEX